W.J. Gage

Gage's health series for primary classes

with special reference to the effects of alcoholic drinks, stiumulants, and

narcotics upon the human system : part 1

W.J. Gage

Gage's health series for primary classes
with special reference to the effects of alcoholic drinks, stiumulants, and narcotics upon the human system : part 1

ISBN/EAN: 9783744741682

Printed in Europe, USA, Canada, Australia, Japan

Cover: Foto ©berggeist007 / pixelio.de

More available books at **www.hansebooks.com**

GAGE'S

HEALTH SERIES

FOR PRIMARY CLASSES

WITH SPECIAL REFERENCE TO THE EFFECTS OF ALCOHOLIC DRINKS,
STIMULANTS, AND NARCOTICS UPON THE HUMAN SYSTEM.

PART I.

Authorized for Use in the Schools of Manitoba.
Authorized for Use in the Schools of British Columbia.

THE W. J. GAGE COMPANY (LTD.)
TORONTO.

Contents

CHAPTER I.

JOINTS AND BONES.

LITTLE girls like a jointed doll to play with, because they can bend such a doll

Jointed dolls.

in eight or ten places, make it stand or sit, or can even play that it is walking.

As you study your own bodies to-day, you will find that you have better joints than any dolls that can be bought at a toy shop.

HINGE-JOINTS.

Some of your joints work like the hinges of a door, and these are called hinge-joints.

You can find them in your elbows, knees, fingers, and toes.

How many hinge-joints can you find?

Think how many hinges must be used by the boy who takes off his hat and makes a polite bow to his teacher, when she meets him on the street.

How many hinges do you use in running up-stairs, opening the door, buttoning your coat or your boots, playing ball or digging in your garden?

You see that we use these hinges nearly all the time. We could not do without them.

BALL AND SOCKET JOINTS.

Not all our joints are hinge-joints.

Your shoulder has a joint that lets your

arm swing round and round, as well as move
up and down.

Your hip has another that lets your leg
move in much the same way.

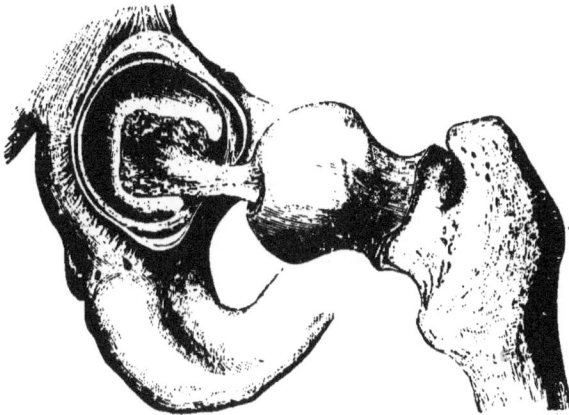

The hip-joint.

This kind of joint is the round end or ball
of a long bone, which moves in a hole, called
a socket.

Your joints do not creak or get out of order, as those of doors and gates sometimes
do. A soft, smooth fluid, much like the white
of an egg, keeps them moist and makes them
work easily.

BONES.

What parts of our bodies are jointed together so nicely? Our bones.

How many bones have we?

If you should count all your bones, you
would find that each of you has about two
hundred.

Some are large; and some, very small.

There are long bones in your legs and
arms, and many short ones in your fingers
and toes. The backbone is called the spine.

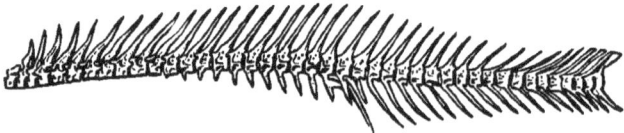

Backbone of a fish.

If you look at the backbone of a fish, you
can see that it is made up of many little
bones. Your own spine is formed in much
the same way, of twenty-four small bones.
An elastic cushion of gristle (grĭs'l) fits nicely
in between each little bone and the next.

When you bend, these cushions are pressed
together on one side and stretched on the

other. They settle back into their first shape, as soon as you stand straight again.

If you ever rode in a wheelbarrow, or a cart without springs, you know what a jolting it gave you. These little spring cushions keep you from being shaken even more severely every time you move.

Twenty-four ribs, twelve on each side, curve around from the spine to the front, or breast, bone. (*See page 38.*)

They are so covered with flesh that perhaps you can not feel and count them; but they are there.

Then you have two flat shoulder-blades, and two collar-bones that almost meet in front, just where your collar fastens.

Of what are the bones made?

Take two little bones, such as those from the legs or wings of a chicken, put one of them into the fire, when it is not very hot, and leave it there two or three hours. Soak the other bone in some weak muriatic (mŏ rĭ ăt'ĭk) acid. This acid can be bought of any druggist.

You will have to be careful in taking the

bone out of the fire, for it is all ready to break. If you strike it a quick blow, it will crumble to dust. This dust we call lime, and it is very much like the lime from which the mason makes mortar.

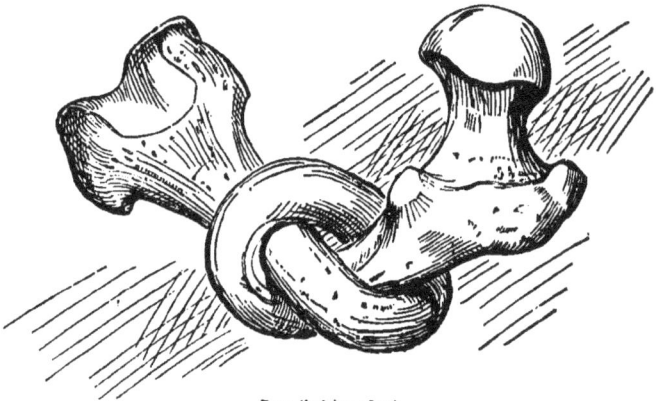

Bone tied in a knot.

The acid has taken the lime **from the** other bone, so only the part which **is not** lime is left. You will be surprised to see how easily it will bend. You can twist it and tie it into a knot; but it will not easily break.

You have seen gristle in meat. This soft part of the bone is gristle.

Children's bones have more gristle than those of older people; so children's bones bend easily.

I know a lady who has one leg shorter than the other. This makes her lame, and she has to wear a boot with iron supports three or four inches high, in order to walk at all.

One day she told me how she became lame.

"I remember," she said, "when I was between three and four years old, sitting one day in my high chair at the table, and twisting one foot under the little step of the chair. The next morning I felt lame; but nobody could tell what was the matter. At last, the doctors found out that the trouble all came from that twist. It had gone too far to be cured. Before I had this boot, I could only walk with a crutch."

CARE OF THE SPINE.

Because the spine is made of little bones with cushions between them, it bends easily, and children sometimes bend it more than they ought.

If you lean over your book or your writing or any other work, the elastic cushions may get so pressed on the inner edge that they do not easily spring back into shape. In this way, you may grow round-shouldered or hump-backed.

This bending over, also cramps the lungs, so that they do not have all the room they need for breathing. While you are young, your bones are easily bent. One shoulder or one hip gets higher than the other, if you stand unevenly. This is more serious, because you are growing, and you may grow crooked before you know it.

Now that you know how soft your bones are, and how easily they bend, you will surely be careful to sit and stand erect. Do not twist your legs, or arms, or shoulders; for you want to grow into straight and graceful men and women, instead of being round-shouldered, or hump-backed, or lame, all your lives.

When people are old, their bones contain more lime, and, therefore, break more easily.

You should be kindly helpful to old peo-

ar writ-
:ushions
ge that
) shape.
)uldered

e lungs,
)m they
young,
ulder or
, if you
because
crooked

ar bones
ou will
:ct. Do
oulders;
d grace-
round-
all your

contain
easily.
)ld peo-

ple, so that they may not fall, and possibly break their bones.

CARE OF THE FEET.

Healthy children are always out-growing their shoes, and sometimes faster than they wear them out. Tight shoes cause corns and in-growing nails and other sore places on the feet. All of these are very hard to get rid of. No one should wear a shoe that pinches or hurts the foot.

OUGHT A BOY TO USE TOBACCO?

Perhaps some boy will say: "Grown people are always telling us, 'this will do for men, but it is not good for boys.'"

Tobacco is not good for men; but there is a very good reason why it is worse for boys.

If you were going to build a house, would it be wise for you to put into the stone-work of the cellar, something that would make it less strong?

Something into the brick-work or the mortar, the wood-work or the nails, the

walls or the chimneys, that would make
them weak and tottering, instead of strong
and steady?

It would be bad enough if you should
repair your house with poor materials; but
surely it must be built in the first place
with the best you can get.

You will soon learn that boys and girls
are building their bodies, day after day, until
at last they reach full size.

Afterward, they must be repaired as fast
as they wear out.

It would be foolish to build any part in
a way to make it weaker than need be.

Wise doctors have said that the boy who
uses tobacco while he is growing, makes
every part of his body less strong than it
otherwise would be. Even his bones will
not grow so well.

Boys who smoke can not become such
large, fine-looking men as they would if
they did not smoke.

Cigarettes are small, but they are very
poisonous. Chewing tobacco is a worse and
more filthy habit even than smoking. The

frequent spitting it causes is disgusting to others and hurts the health of the chewer. Tobacco in any form is a great enemy to youth. It stunts the growth, hurts the mind, and cripples in every way the boy or girl who uses it.

Not that it does all this to every youth who smokes, but it is always true that no boy of seven to fourteen can begin to smoke or chew and have so fine a body and mind when he is twenty-one years old as he would have had if he had never used tobacco. If you want to be strong and well men and women, do not use tobacco in any form.

REVIEW QUESTIONS.

1. What two kinds of joints have you?
2. Describe each kind.
3. Find as many of each kind as you can.
4. How are the joints kept moist?
5. How many bones are there in your whole body?
6. Count the bones in your hand.
7. Of how many bones is your spine made?
8. Why could you not use it so well if it were all in one piece?
9. What is the use of the little cushions between the bones of the spine?
10. How many ribs have you?
11. Where are they?
12. Where are the shoulder-blades?
13. Where are the collar-bones?

14. What are bones made of?
15. How can we show this?
16. What is the difference between the bones of children and the bones of old people?
17. Why do children's bones bend easily?
18. Tell the story of the lame lady.
19. What does this story teach you?
20. What happens if you lean over your desk or work?
21. How will this position injure your lungs?
22. What other bones may be injured by wrong positions?
23. Why do old people's bones break easily?
24. How should the feet be cared for?
25. How does tobacco affect the bones?
26. What do doctors say of its use?
27. What is said about cigarettes?
28. What about chewing tobacco?
29. To whom is tobacco a great enemy? Why?
30. What is always true of its use by youth?

CHAPTER II.

MUSCLES.

WHAT makes the limbs move?

You have to take hold of the door to move it back and forth; but you need not take hold of your arm to move that.

What makes it move?

Sometimes a door or gate is made to shut itself, if you leave it open.

This can be done by means of a wide rubber strap, one end of which is fastened to the frame of the door near the hinge, and the other end to the door, out near its edge.

When we push open the door, the rubber strap is stretched; but as soon as we have passed through, the strap tightens, draws the door back, and shuts it.

If you stretch out your right arm, and clasp the upper part tightly with your left hand, then work the elbow joint strongly back and forth, you can feel something under your hand draw up, and then lengthen out again, each time you bend the joint.

What you feel, is a muscle (mŭs'sl), and it works your joints very much as the rubber strap works the hinge of the door.

One end of the muscle is fastened to the bone just below the elbow joint; and the other end, higher up above the joint.

When it tightens, or contracts, as we say, it bends the joint. When the arm is straightened, the muscle returns to its first shape.

There is another muscle on the outside of the arm which stretches when this one shortens, and so helps the working of the joint.

Every joint has two or more muscles of its own to work it.

Think how many there must be in our fingers!

If we should undertake to count all the muscles that move our whole bodies, it would need more counting than some of you could do.

TENDONS. 23

TENDONS.

You can see muscles on the dinner table; for they are only lean meat.

Tendons of the hand.

They are fastened to the bones by strong cords, called tendons (těn'dŏnz). These tendons can be seen in the leg of a chicken or turkey They sometimes hold the meat so firmly that it is hard for you to get it off. When you next try to pick a "drum-stick," remember that you are eating the strong muscles by which the chicken or turkey moved his legs as he walked about the yard. The parts that have the most work to do, need the strongest muscles.

Did you ever see the swallows flying about the eaves of a barn?

Do they have very stout legs? No! They

have very small legs and feet, because they
do not need to walk. They need to fly.

The muscles that move the wings are
fastened to the breast. These breast muscles
of the swallow must be large and strong.

EXERCISE OF THE MUSCLES.

People who work hard with any part of
the body make the muscles of that part very
strong.

The blacksmith has big, strong muscles
in his arms because he uses them so much.

You are using your muscles every day,
and this helps them to grow.

Once I saw a little girl who had been
very sick. She had to lie in bed for many
weeks. Before her sickness she had plenty
of stout muscles in her arms and legs and
was running about the house from morning
till night, carrying her big doll in her arms.

After her sickness, she could hardly walk
ten steps, and would rather sit and look at
her playthings than try to lift them. She
had to make new muscles as fast as possible.

Running, coasting, games of ball, and all

brisk play and work, help to make strong
muscles.

Idle habits make weak muscles. So idle-
ness is an enemy to the muscles.

There is another enemy to the muscles
about which I must tell you.

WHAT ALCOHOL WILL DO TO THE MUSCLES.

Muscles are lean meat. Fat meat could
not work your joints for you as the muscles
do. Alcohol often changes a part of the
muscles to fat, and so takes away a part of
their strength. In this way, people often
grow very fleshy from drinking beer, because
it contains alcohol, as you will soon learn.
But they can not work any better on account of having his fat. They are not really
any stronger for it.

REVIEW QUESTIONS

1. How are the joints moved?
2. Where are the muscles in your arms, which help you to move
   your elbows?
3. Show why joints must have muscles.
4. What do we call the muscles of the lower animals?
5. What fasten the muscles to the bones?

6. Why do chickens and turkeys need strong muscles in their legs?
7. Why do swallows need strong breast muscles?
8. What makes the muscles of the blacksmith's arm so strong?
9. What will make your muscles strong?
10. What will make them weak?
11. What does alcohol often do to the muscles?
12. Can fatty muscles work well?
13. Why does not drinking beer make one stronger?

CHAPTER III.

NERVES.

HOW do the muscles know when to move? You have all seen the telegraph wires, by which messages are sent from one town to another, all over the country.

You are too young to understand how this is done, but you each have something inside of you, by which you are sending messages almost every minute while you are awake.

We will try to learn a little about its wonderful way of working.

In your head is your brain. It is the part of you which thinks.

As you would be very badly off if you could not think, the brain is your most precious part, and you have a strong box made of bone to keep it in.

NERVES.

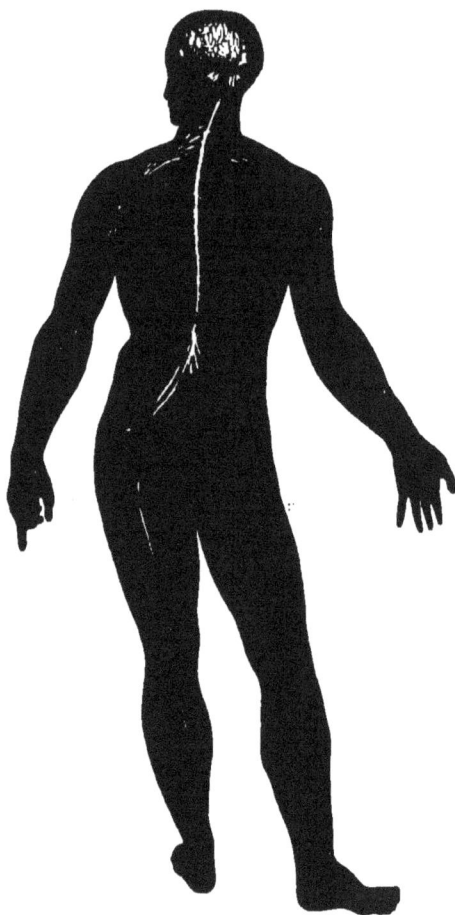

Diagram of the nervous system.

We will call the brain the central telegraph office. Little white cords, called nerves,
connect the brain with the rest of the body.

A large cord called the spinal cord, lies
safely in a bony case made by the spine, and
many nerves branch off from this.

If you put your finger on a hot stove, in
an instant a message goes on the nerve telegraph to the brain. It tells that wise thinking part that your finger will burn, if it
stays on the stove.

In another instant, the brain sends back
a message to the muscles that move that
finger, saying: "Contract quickly, bend the
joint, and take that poor finger away so
that it will not be burned."

You can hardly believe that there was
time for all this sending of messages; for as
soon as you felt the hot stove, you pulled
your finger away. But you really could not
have pulled it away, unless the brain had
sent word to the muscles to do it.

Now, you know what we mean when we
say, "As quick as thought." Surely nothing could be quicker.

You see that the brain has a great deal
of work to do, for it has to send so many or-
ders.

There are some muscles which are moving quietly and steadily all the time, though
we take no notice of the motion.

You do not have to think about breathing, and yet the muscles work all the time,
moving your chest.

If we had to think about it every time
we breathed, we should have no time to
think of any thing else.

There is one part of the brain that takes
care of such work for us. It sends the messages about breathing, and keeps the breathing muscles and many other muscles faithfully at work. It does all this without our
needing to know or think about it at all.

Do you begin to see that your body is a
busy work-shop, where many kinds of work
are being done all day and all night?

Although we lie still and sleep in the
night, the breathing must go on, and so must
the work of those other organs that never
stop until we die.

reat deal
many or-

re mov-
, though

breath-
he time,

ry time
time to

at takes
he mes-
breath-
s faith-
out our
all.
dy is a
of work

in the
o must
never

OTHER WORK OF THE NERVES.

The little white nerve-threads lie smoothly side by side, making small white cords. Each kind of message goes on its own thread, so that the messages need never get mixed or confused.

These nerves are very delicate little messengers. They do all the feeling for the whole body, and by means of them we have many pains and many pleasures.

If there was no nerve in your tooth it could not ache. But if there were no nerves in your mouth and tongue, you could not taste your food.

If there were no nerves in your hands, you might cut them and feel no pain. But you could not feel your mother's soft, warm hand, as she laid it on yours.

One of your first duties is the care of yourselves.

Children may say: "My father and mother take care of me." But even while you are young, there are some ways in which no one can take care of you but yourselves. The

older you grow, the more this care will be-
long to you, and to no one else.

Think of the work all the parts of the
body do for us, and how they help us to be
well and happy. Certainly the least we can
do is to take care of them and keep them in
good order.

CARE OF THE BRAIN AND NERVES.

As one part of the brain has to take care of
all the rest of the body, and keep every organ
at work, of course it can never go to sleep
itself. If it did, the heart would stop pumping, the lungs would leave off breathing, all
other work would stop, and the body would
be dead.

But there is another part of the brain
which does the thinking, and this part needs
rest.

When you are asleep, you are not thinking, but you are breathing and other work of
the body is going on.

If the thinking part of the brain does not
have good quiet sleep, it will soon wear out.
A worn-out brain is not easy to repair,

will be-

s of the
us to be
t we can
them in

VES.

e care of
ry organ
to sleep
p pump-
hing, all
ly would

he brain
rt needs

t think-
work of

does not
ear out.

If well cared for, your brain will do the best of work for you for seventy or eighty years without complaining.

The nerves are easily tired out, and they need much rest. They get tired if we do one thing too long at a time; they are rested by a change of work.

IS ALCOHOL GOOD FOR THE NERVES AND THE BRAIN?

Think of the wonderful work the brain is all the time doing for you!

You ought to give it the best of food to keep it in good working order. Any drink that contains alcohol is not a food to make one strong; but is a poison to hurt, and at last to kill.

It injures the brain and nerves so that they can not work well, and send their messages properly. That is why the drunkard does not know what he is about.

Newspapers often tell us about people setting houses on fire; about men who forgot to turn the switch, and so wrecked a railroad train; about men who lay down on the railroad track and were run over by the cars.

Often these stories end with: "The person had been drinking." When the nerves are put to sleep by alcohol, people become careless and do not do their work faithfully; sometimes, they can not even tell the difference between a railroad track and a place of safety. The brain receives no message, or the wrong one, and the person does not know what he is doing.

You may say that all men who drink liquor do not do such terrible things.

That is true. A little alcohol is not so bad as a great deal. But even a little makes the head ache, and hurts the brain and nerves.

A body kept pure and strong is of great service to its owner. There are people who are not drunkards, but who often drink a little liquor. By this means, they slowly poison their bodies.

When sickness comes upon them, they are less able to bear it, and less likely to get well again, than those who have never injured their bodies with alcohol.

When a sick or wounded man is brought

into the hospital, one of the first questions
asked him by the doctor is : "Do you drink ?"

If he answers "Yes!" the next questions
are. "What do you drink?" and "How
much ?"

The answers he gives to these questions,
show the doctor what chance the man has
of getting well.

A man who never drinks liquor will get
well, where a drinking man would surely
die.

TOBACCO AND THE NERVES.

Why does any one wish to use tobacco?

Because many men say that it helps them,
and makes them feel better.

Shall I tell you how it makes them feel
better?

If a man is cold, the tobacco deadens
his nerves so that he does not feel the
cold and does not take pains to make himself
warmer.

If a man is tired, or in trouble, tobacco
will not really rest him or help him out of
his trouble.

It only puts his nerves to sleep and helps him think that he is not tired, and that he does not need to overcome his troubles.

It puts his nerves to sleep very much as alcohol does, and helps him to be contented with what ought not to content him.

A boy who smokes or chews tobacco, is not so good a scholar as if he did not use the poison. He can not remember his lessons so well.

Usually, too, he is not so polite, nor so good a boy as he otherwise would be.

REVIEW QUESTIONS.

1. How do the muscles know when to move?
2. What part of you is it that thinks?
3. What are the nerves?
4. Where is the spinal cord?
5. What message goes to the brain when you put your finger on a hot stove?
6. What message comes back from the brain to the finger?
7. What is meant by "As quick as thought"?
8. Name some of the muscles which work without needing our thought.
9. What keeps them at work?
10. Why do not the nerve messages get mixed and confused?
11. Why could you not feel, if you had no nerves?
12. State some ways in which the nerves give us pain.
13. State some ways in which they give us pleasure.
14. What part of us has the most work to do?

15. How must we keep the brain strong and well?
16. What does alcohol do to the nerves and brain?
17. Why does not a drunken man know what he is about?
18. What causes most of the accidents we read of?
19. Why could not the man who had been drinking tell the difference between a railroad track and a place of safety?
20. How does the frequent drinking of a little liquor affect the body?
21. How does sickness affect people who often drink these liquors?
22. When a man is taken to the hospital, what questions does the doctor ask?
23. What depends upon his answers?
24. Why do many men use tobacco?
25. How does it make them feel better?
26. Does it really help a person who uses it?
27. Does tobacco help a boy to be a good scholar?
28. How does it affect his manners?

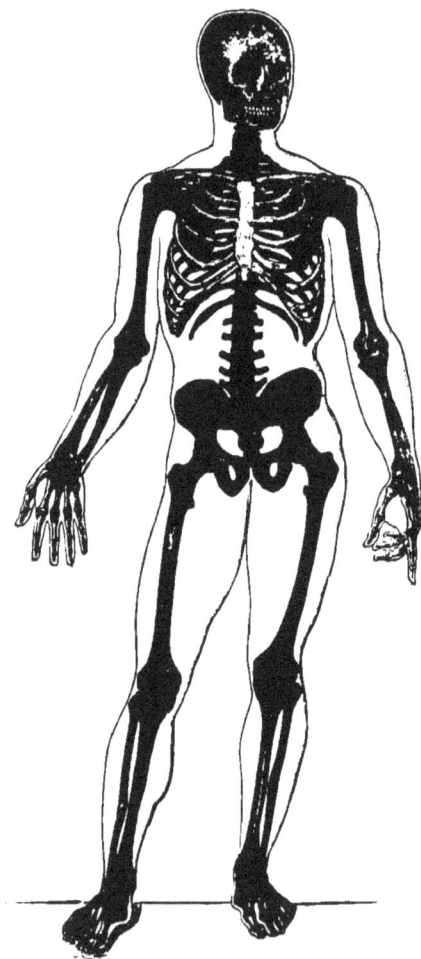

Bones of the human body.

CHAPTER IV.

WHAT IS ALCOHOL?

RIPE grapes are full of juice.

This juice is mostly water, sweetened with a sugar of its own. It is flavored with something which makes us know, the moment we taste it, that it is grape-juice, and not cherry-juice or plum-juice.

Apples also contain water, sugar, and apple flavor; and cherries contain water, sugar, and cherry flavor. The same is true of other fruits. They all, when ripe, have the water and the sugar; and each has a flavor of its own.

Ripe grapes are sometimes gathered and put into great tubs called vats. In these the juice is squeezed out.

In some countries, this squeezing is done by bare-footed men who jump into the vats and press the grapes with their feet.

The grape-juice is then drawn off from the
skins and seeds and left standing in a warm
place.

Bubbles soon begin to rise and cover the
top of it with froth. The juice is all in motion.

Picking grapes and making wine.

If the cook had wished to use this grape-
juice to make jelly, she would say: "Now, I
can not make my grape-jelly, for the grape-
juice is spoiled."

WHAT IS THIS CHANGE IN THE GRAPE-JUICE?

The sugar in the grape-juice is changing into something else. It is turning into alcohol and a gas* that moves about in little bubbles in the liquid, and rising to the top, goes off into the air. The alcohol is a thin liquid which, mixed with the water, remains in the grape-juice.

The sugar is gone; alcohol and the bubbles of gas are left in its place.

This alcohol is a liquid poison. A little of it will harm any one who drinks it; much of it would kill the drinker.

Ripe grapes are good food; but grape-juice, when its sugar has turned to alcohol, is not a safe drink for any one. It is poisoned by the alcohol.

WINE.

This changed grape-juice is called wine. It is partly water, partly alcohol, and it still has the grape flavor in it.

---

* This gas is called car bon'ic acid gas.

Wine is also made from currants, elderberries, and other fruits, in very much the same way as from grapes.

People sometimes make it at home from the fruits that grow in their own gardens, and think there is no alcohol in it, because they do not put any in.

But you know that the alcohol is made in the fruit-juice itself by the change of the sugar into alcohol and the gas.

It is the nature of alcohol to make the person who takes a little of it, in wine, or any other drink, want more and more alcohol. When one goes on, thus taking more and more of the drinks that contain alcohol, he is called a drunkard.

In this way wine has made many d.unk-

ards. Alcohol hurts both the body and mind. It changes the person who drinks it. It will make a good and kind person cruel and bad; and will make a bad person worse.

Not every one who takes wine becomes a drunkard, but you are not sure that you will not, if you drink it.

CIDER.

Cider is made from apples. In a few hours after the juice is pressed out of the apples, if it is left open to the air the sugar begins to change.

Like the sugar in the grape, it changes into alcohol and bubbles of gas.

At first, there is but little alcohol in cider, but a little of this poison is dangerous.

More alcohol is all the time forming until in ten cups of cider there may be one cup of alcohol. Cider often makes its drinkers ill-tempered and cross.

Cider and wine will turn into vinegar if left in a warm place long enough.

REVIEW QUESTIONS.

1. What two things are in all fruit-juices?
2. How can we tell the juice of grapes from that of plums?
3. How can we tell the juice of apples from that of cherries?
4. What is often done with ripe grapes?
5. What happens after the grape-juice has stood a short time?
6. Why would the changed grape-juice not be good to use in making jelly?
7. Into what is the sugar in the juice changed?
8. What becomes of the gas?
9. What becomes of the alcohol?
10. What is gone and what left?
11. What is alcohol?
12. What does alcohol do to those who drink it?
13. When are grapes good food?
14. When is grape-juice not a safe drink?
15. Why?
16. What is this changed grape-juice called?
17. What is wine?
18. From what is wine made?
19. What do people sometimes think of home-made wines?
20. How can alcohol be there when none has been put into it?
21. What does alcohol make the person who takes it want?
22. What is such a one called?
23. What has wine done to many persons?
24. What does alcohol hurt?
25. How does it change a person?
26. Are you sure you will not become a drunkard if you drink wine?
27. Why should you not drink it?
28. What is cider made from?
29. What soon happens to apple-juice?
30. How may vinegar be made?

CHAPTER V.

BEER.

ALCOHOL is often made from grains as well as from fruit. The grain has starch instead of sugar.

If the starch in your mother's starch-box at home should be changed into sugar, you would think it a very strange thing.

Every year, in the spring-time, many thousand pounds of starch are changed into sugar in a hidden, quiet way, so that most of us think nothing about it.

STARCH AND SUGAR.

All kinds of grain are full of starch.

If you plant them in the ground, where they are kept moist and warm, they begin to sprout and grow, to send little roots down into the earth, and little stems up into the sunshine.

These little roots and stems must be fed
with sugar; thus, in a wise way, which is too
wonderful for you to understand, as soon
as the seed begins to sprout,
its starch begins to turn into
sugar.

If you should chew two
grains of wheat, one before
sprouting and one after, you
could tell by the taste that
this is true.

Barley is a kind of grain
from which the brewer
makes beer.

He must first turn
its starch into sugar, so
he begins by sprouting
his grain.

Of course he does not
plant it in the ground,
because it would need
to be quickly dug up again.

He keeps it warm and moist in a place
where he can watch it, and stop the sprouting just in time to save the sugar, before it

st be fed
ich is too
as soon
) sprout,
urn into

ew two
before
er, you
to that

f grain
brewer

it turn
ugar, so
outing

oes not
round,
need

place
prout-
ore it

is used to feed the root and stem. This sprouted grain is called malt.

The brewer soaks it in plenty of water, because the grain has not water in itself, as the grape has.

He puts in some yeast to help start the work of changing the sugar into gas* and alcohol.

Sometimes hops are also put in, to give it a bitter taste.

The brewer watches to see the bubbles of gas that tell, as plainly as words could, that sugar is going and alcohol is coming.

When the work is finished, the barley has been made into beer.

It might have been ground and made into barley-cakes, or into pearl barley to thicken our soups, and then it would have been good food. Now, it is a drink containing alcohol, and alcohol is a poison.

You should not drink beer, because there is alcohol in it.

Two boys of the same age begin school

* Car bon' ic acid gas.

together. One of them drinks wine, cider, and beer. The other never allows these drinks to pass his lips. These boys soon become very different from each other, because one is poisoning his body and mind with alcohol, and the other is not.

A man wants a good, steady boy to work for him. Which of these two do you think he will select? A few years later, a young man is wanted who can be trusted with the care of an engine or a bank. It is a good chance. Which of these young men will be more likely to get it?

REVIEW QUESTIONS.

1. Is there sugar in grain?
2. What is in the grain that can be turned into sugar?
3. What can you do to a seed that will make its starch turn into sugar?
4. What does the brewer do to the barley to make its starch turn into sugar?
5. What is malt?
6. What does the brewer put into the malt to start the working?
7. What gives the bitter taste to beer?
8. How does the brewer know when sugar begins to go and alcohol to come?
9. Why does he want the starch turned to sugar?
10. Is barley good for food?
11. Why is beer not good for food?
12. Why should you not drink it?
13. Why did the two boys of the same age, at the same school, become so unlike?
14. Which will have the best chance in life?

, cider,

these

oon be-

because

d with

o work

think

young

ith the

a good

will be

turn into

arch turn

working?

and alco-

chool, be-

CHAPTER VI.

DISTILLING.

DISTILLING (dis til'ing) may be a new word to you, but you can easily learn its meaning.

You have all seen distilling going on in the kitchen at home, many a time. When the water in the tea-kettle is boiling, what comes out at the nose? Steam.

What is steam?

You can find out what it is by catching some of it on a cold plate, or tin cover. As soon as it touches any thing cold, it turns into drops of water.

When we boil water and turn it into steam, and then turn the steam back into water, we have distilled the water. We say vapor instead of steam, when we talk about the boiling of alcohol.

It takes less heat to turn alcohol to vapor

than to turn water to steam; so, if we put
over the fire some liquid that contains alcohol, and begin to collect the vapor as it rises,
we shall get alcohol first, and then water.

But the alcohol will not be pure alcohol;
it will be part water, because it is so ready
to mix with water that it has to be distilled
many times to be pure.

But each time it is distilled, it will become stronger, because there is a little more
alcohol and a little less water.

In this way, brandy, rum, whiskey, and
gin are distilled, from wine, cider, and the
liquors which have been made from corn,
rye, or barley.

The cider, wine, and beer had but little
alcohol in them. The brandy, rum, whiskey,
and gin are nearly one-half alcohol.

A glass of strong liquor which has been
made by distilling, will injure any one more,
and quicker, than a glass of cider, **wine, or**
beer.

But a cider, wine, or beer-drinker often
drinks so much more of the weaker liquor,
that he gets a great deal of alcohol. People

we put
ns alco-
it rises,
water.
alcohol;
so ready
distilled

will be-
:lo more

:ey, and
and the
m corn,

ut little
whiskey,

.ias been
io more,
wine, or

er often
r liquor,
People

are often made drunkards by drinking cider
or beer. The more poison, the more danger.

REVIEW QUESTIONS.

1. Where have you ever seen distilling going on?
2. How can you distill water?
3. How can men separate alcohol from wine or from any other liquor that contains it?
4. Why will not this be pure alcohol?
5. How is a liquor made stronger?
6. Name some of the distilled liquors.
7. How are they made?
8. How much of them is alcohol?
9. Which is the most harmful—the distilled liquor, or beer, wine, or cider?
10. Why does the wine, cider, or beer drinker often get as much alcohol?

CHAPTER VII.

ALCOHOL.

ALCOHOL looks like water, but it is not at all like water.

Alcohol will take fire, and burn if a lighted match is held near it; but you know that water will not burn.

When alcohol burns, the color of the flame is blue. It does not give much light: it makes no smoke or soot; but it does give a great deal of heat.

A little dead tree-toad was once put into a bottle of alcohol. It was years ago, but the tree-toad is there still, looking just as it did the first day it was put in. What has kept it so?

It is the alcohol. The tree-toad would have soon decayed if it had been put into water. So you see that alcohol keeps dead bodies from decaying.

Pure alcohol is not often used as a drink. People who take beer, wine, and cider get a little alcohol with each drink. Those who drink brandy, rum, whiskey, or gin, get more alcohol, because those liquors are nearly one half alcohol.

You may wonder that people wish to use such poisonous drinks at all. But alcohol is a deceiver. It often cheats the man who takes a little, into thinking it will be good for him to take more.

Sometimes the appetite which begs so hard for the poison, is formed in childhood. If you eat wine-jelly, or wine-sauce, you may learn to like the taste of alcohol and thus easily begin to drink some weak liquor.

The more the drinker takes, the more he often wants, and thus he goes on from drinking cider, wine, or beer, to drinking whiskey, brandy, or rum. Thus drunkards are made.

People who are in the habit of taking drinks which contain alcohol, often care more for them than for any thing else, even when they know they are being ruined by them.

REVIEW QUESTIONS.

1. How does alcohol look ?
2. How does alcohol burn ?
3. What will alcohol do to a dead body ?
4. What drinks contain a little alcohol ?
5. What drinks are about one half alcohol ?
6. How does alcohol cheat people ?
7. When is the appetite sometimes formed ?
8. Why should you not eat wine-sauce or wine-jelly ?
9. How are drunkards made ?

CHAPTER VIII.

TOBACCO.

A FARMER who had been in the habit of planting his fields with corn, wheat, and potat once made up his mind to plant tobacco instead.

Let us see whether he did any good to the world by the change.

The tobacco plants grew up as tall as a little boy or girl, and spread out broad, green leaves.

By and by he pulled the stalks, and dried the leaves. Some of them he pressed into cakes of tobacco; some he rolled into cigars; and some he ground into snuff.

If you ask what tobacco is good for, the best answer will be, to tell you what it will do to a man or boy who uses it, and then let you answer the question for yourselves.

Tobacco contains something called nico-

tine (nĭk'o ŭ). This is a strong poison. One
drop of it is enough to kill a dog. In one
cigar there is enough, if taken pure, to kill
two men.

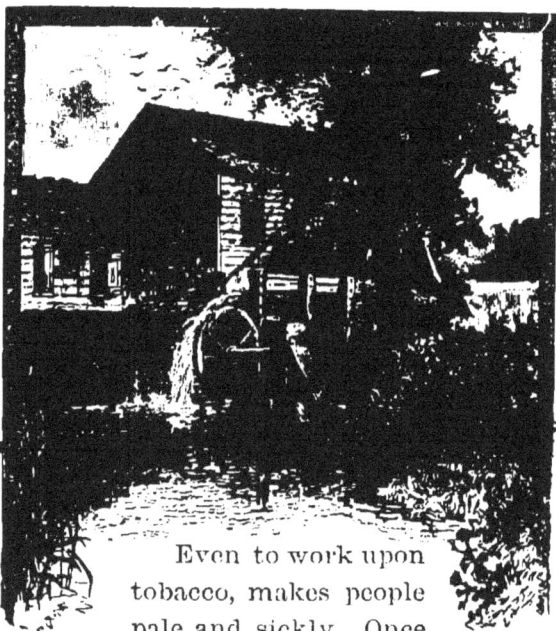

Even to work upon
tobacco, makes people
pale and sickly. Once
I went into a snuff mill, and the man who
had the care of it showed me how the work
was done.

The mill stood in a pretty place, beside a little stream which turned the mill-wheel. Tall trees bent over it, and a fresh breeze was blowing through the open windows. Yet the smell of the tobacco was so strong that I had to go to the door many times, for a breath of pure air.

I asked the man if it did not make him sick to work there.

He said: "It made me very sick for the first few weeks. Then I began to get used to it, and now I don't mind it."

He was like the boys who try to learn to smoke. It almost always makes them sick at first; but they think it will be manly to keep on. At last, they get used to it.

The sickness is really the way in which the boy's body is trying to say to him: "There is danger here; you are playing with poison. Let me stop you before great harm is done."

Perhaps you will say: "I have seen men smoke cigars, even four or five in a day, and it didn't kill them."

It did not kill them, because they did

not swallow the nicotine. They only drew
in a little with the breath. But taking a
little poison in this way, day after day, can
not be safe, or really helpful to any one.

REVIEW QUESTIONS.

1. What did the farmer plant instead of corn, wheat, and potatoes?
2. What was done with the tobacco leaves?
3. What is the name of the poison which is in tobacco?
4. How much of it is needed to kill a dog?
5. What harm can the nicotine in one cigar do, if taken pure?
6. Tell the story of the visit to the snuff mill.
7. Why are boys made sick by their first use of tobacco?
8. Why does not smoking a cigar kill a man?
9. What is said about a little poison?

CHAPTER IX.

OPIUM.

ALCOHOL and tobacco are called narcotics (nar kŏt′iks). This means that they have the power of putting the nerves to sleep. Opium (ō′pĭ ŭm) is another narcotic.

It is a poison made from the juice of poppies, and is used in medicines.

Opium is put into soothing-syrups (sĭr′ŭps), and these are sometimes given to babies to keep them from crying. They do this by injuring the tender nerves and poisoning the little body.

How can any one give a baby opium to save taking patient care of it?

Surely the mothers would not do it, if they knew that this soothing-syrup that appears like a friend, coming to quiet and comfort the baby, is really an enemy.

Sometimes, a child no older than some of

you are, is left at home with the care of a
baby brother or sister; so it is best that you

Don't give soothing-syrup to children.

should know about this dangerous enemy,
and never be tempted to quiet the baby by
giving him a poison, instead of taking your
best and kindest care of him.

REVIEW QUESTIONS.

1. What is a narcotic?
2. Name three narcotics.
3. **From what is opium made?**
4. For what is it used?
5. Why is soothing-syrup dangerous?

CHAPTER X.

WHAT ARE ORGANS?

AN organ is a part of the body which has some special work to do. The eye is the organ of sight. The stomach (stŭm'ăk) is an organ which takes care of the food we eat.

THE TEETH.

Your teeth do not look alike, since they

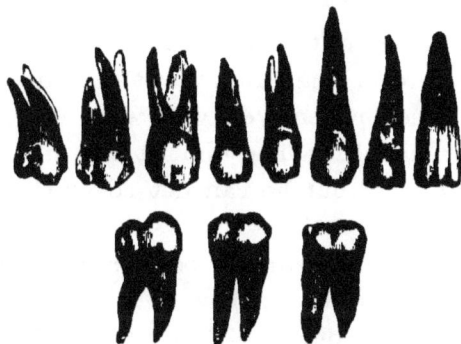

Different kinds of teeth.

must do different kinds of work. The front
ones cut, the back ones grind.

They are made of a kind of bone covered
with a hard smooth enamel (ĕn ăm′el). If the
enamel is broken, the teeth soon decay and
ache, for each tooth is furnished with a nerve
that very quickly feels pain.

CARE OF THE TEETH.

Cracking nuts with the teeth, or even
biting thread, is apt to break the enamel; and
when once broken, you will wish in vain to
have it mended. The dentist can fill a hole
in the tooth; but he can not cover the tooth
with new enamel.

Bits of food should be carefully picked
from between the teeth with a tooth-pick of
quill or wood, never with a pin or other hard
and sharp thing which might break the
enamel.

The teeth must also be well brushed.
Nothing but perfect cleanliness will keep
them in good order. Always brush them
before breakfast. Your breakfast will taste
all the better for it. Brush them at night

he front

covered
If the
:ay and
a nerve

or even
iel; and
vain to
a hole
e tooth

picked
pick of
er hard
ak the

rushed.
.l keep
them
l taste
night

before you go to bed, lest some food should be decaying in your mouth during the night.

Take care of these cutters and grinders, that they may not decay, and so be unable to do their work well.

THE CHEST AND ABDOMEN.

You have learned about the twenty-four little bones in the spine, and the ribs that curve around from the spine to the front, or breast-bone.

These bones, with the shoulder-blades and the collar-bones, form a bony case or box.

In it are some of the most useful organs of the body.

This box is divided across the middle by a strong muscle, so that we may say it is two stories high.

The upper room is called the chest; the lower one, the abdomen (ăb dō'mĕn).

In the chest, are the heart and the lungs.

In the abdomen, are the stomach, the liver, and some other organs.

THE STOMACH.

The stomach is a strong bag, as wonderful a bag as could be made, you will say, when I tell you what it can do.

The outside is made of muscles; the lining prepares a juice called gastric (găs′trĭk) juice, and keeps it always ready for use.

Now, what would you think if a man could put into a bag, beef, and apples, and potatoes, and bread and milk, and sugar, and salt, tie up the bag and lay it away on a shelf for a few hours, and then show you that the beef had disappeared, so had the apples, so had the potatoes, the bread and milk, sugar, and salt, and the bag was filled only with a thin, grayish fluid? Would you not call it a magical bag?

Now, your stomach and mine are just such magical bags.

We put in our breakfasts, dinners, and suppers; and, after a few hours, they are changed. The gastric juice has been mixed with them. The strong muscles that form the outside of the stomach have been squeezing the food, rolling it about, and mixing it together, until it has all been changed to a thin, grayish fluid.

HOW DOES ANYBODY KNOW THIS?

A soldier was once shot in the side in such a way that when the wound healed, it left an opening with a piece of loose skin over it, like a little door leading into his stomach.

A doctor who wished to learn about the stomach, hired him for a servant and used to study him every day.

He would push aside the little flap of skin and put into the stomach any kind of food that he pleased, and then watch to see what happened to it.

In this way, he learned a great deal and wrote it down, so that other people might know, too. In other ways, also, which it would take too long to tell you here, doctors have learned how these magical food-bags take care of our food.

WHY DOES THE FOOD NEED TO BE CHANGED?

Your mamma tells you sometimes at breakfast that you must eat oat-meal and milk to make you grow into a big man or woman.

Did you ever wonder what part of you is made of oat-meal, or what part of milk?

That stout little arm does not look like oat-meal; those rosy cheeks do not look like milk.

If our food is to make stout arms and rosy cheeks, strong bodies and busy brains, it must first be¹ changed into a form in which it can get to each part and feed it.

When the food in the stomach is mixed and prepared, it is ready to be sent through the body; some is carried to the bones, some to the muscles, some to the nerves and brain, some to the skin, and some even to the finger nails, the hair, and the eyes. Each part needs to be fed in order to grow.

WHY DO PEOPLE WHO ARE NOT GROWING NEED FOOD?

Children need each day to make larger and larger bones, larger muscles, and a larger skin to cover the larger body.

Every day, each part is also wearing out

a little, and needing to be mended by some new food. People who have grown up, need their food for this work of mending.

CARE OF THE STOMACH.

One way to take care of the stomach is to give it only its own work to do. The teeth must first do their work faithfully.

The stomach must have rest, too. I have seen some children who want to make their poor stomachs work all the time. They are always eating apples, or candy, or something, so that their stomachs have no chance to rest. If the stomach does not rest, it will wear out the same as a machine would.

The stomach can not work well, unless it is quite warm. If a person pours ice-water into his stomach as he eats, just as the food is beginning to change into the gray fluid of which you have learned, the work stops until the stomach gets warm again.

ALCOHOL AND THE STOMACH.

You remember about the man who had the little door to his stomach. Sometimes,

man or

of you is

nilk?

ook like

.ook like

rms and

r brains,

form in

'eed it.

is mixed

through

.es, some

id brain,

ie finger

.rt needs

ROWING

e larger

a larger

ing out

the doctor put in wine, cider, brandy, or
some drink that contained alcohol, to see
what it would do. It was carried away very
quickly; but during the little time it stayed,
it did nothing but harm.

It injured the gastric juice, so that it
could not mix with the food.

If the doctor had put in more alcohol,
day after day, as one does who drinks liquor,
sores would perhaps have come on the delicate lining of the stomach. Sometimes the
stomach is so hurt by alcohol, that the
drinker dies. If the stomach can not do its
work well, the whole body must suffer from
want of the good food it needs.*

TOBACCO AND THE MOUTH.

The saliva in the mouth helps to prepare
the food, before it goes into the stomach. Tobacco makes the mouth very dry, and more
saliva has to flow out to moisten it.

But tobacco juice is mixed with the saliva, and that must not be swallowed. It

---

* The food is partly prepared by the liver and some other
organs.

must be spit out, and with it is sent the saliva that was needed to help prepare the food.

Tobacco discolors the teeth, makes bad sores in the mouth, and oft n causes a disease of the throat.

You can tell where some people have been, by the neatness and comfort they leave after them.

You can tell where the tobacco-user has been, by the dirty floor, and street, and the air made unfit to breathe, because of the smoke and strong, bad smell of old tobacco from his pipe and cigar and from his breath and clothes.

REVIEW QUESTIONS.

1. What are organs?
2. What work do the front teeth do? the back teeth?
3. What are the teeth made of?
4. What causes the toothache?
5. How is the enamel often broken?
6. Why should a tooth-pick be used?
7. Why should the teeth be well brushed?
8. When should they be brushed?
9. What bones form a case or box?
10. What is the upper room of this box called? the lower room?

11. What organs are in the chest? the abdomen?
12. What is the stomach?
13. What does its lining do?
14. What do the stomach and the gastric juice do to the food we have eaten?
15. How did anybody find out what the stomach could do?
16. Why must all the food we eat be changed?
17. Why do you need food?
18. Why do people who are not growing need food?
19. What does alcohol do to the gastric juice? to the stomach?
20. What is the use of the saliva?
21. How does the habit of spitting injure a person?
22. How does tobacco affect the teeth? the mouth?
23. How does the tobacco-user annoy other p   ....?

CHAPTER XI.

WHAT DOES THE BODY NEED FOR FOOD?

NOW that you know how the body is fed, you must next learn what to feed it with; and what each part needs to make it grow and to keep it strong and well.

WATER.

A large part of your body is made of water. So you need, of course, to drink water, and to have it used in preparing your food.

Water comes from the clouds, and is stored up in cisterns or in springs in the ground. From these pipes are laid to lead the water to our houses.

Sometimes, men dig down until they reach a spring, and so make a well from which they can pump the water, or dip it out with a bucket.

Water that has been standing in lead pipes, may have some of the lead mixed with it. Such water would be very likely to poison you, if you drank it.

Impurities are almost sure to soak into a well if it is near a drain or a stable.

If you drink the water from such a well, you may be made very sick by it. It is better to go thirsty, until you can get good water.

A sufficient quantity of pure water to drink is just as important for us, as good food to eat.

We could not drink all the water that our bodies need. We take a large part of it in our food, in fruits and vegetables, and even in beefsteak and bread.

LIME.

Bones need lime. You remember the bone that was nothing but crumbling lime after it had been in the fire.

Where shall we get lime for our bones?

We can not eat lime; but the grass and the grains take it out of the earth. Then

ead pipes,
with it.
so poison

k into a

in a well,
It is bet-
get good

water to
as good

that our
it in our
even in

the bone
me after

one,
rass and
hen

the cows eat the grass and turn it into milk, and in the milk we drink, we get some of the lime to feed our bones.

Lime being prepared for our use.

In the same way, the grain growing in the field takes up lime and other things that we need, but could not eat for ourselves. The lime that thus becomes a part of the grain, we get in our bread, oat-meal porridge, and other foods.

SALT.

Animals need salt, as children who live in the country know very well. They have seen how eagerly the cows and the sheep lick up the salt that the farmer gives them.

Even wild cattle and buffaloes seek out places where there are salt springs, and go in great herds to get the salt.

We, too, need some salt mixed with our food. If we did not put it in, either when cooking, or afterward, we should still get a little in the food itself.

FLESH-MAKING FOODS.

Muscles are lean meat, that is flesh; so muscles need flesh-making foods. These are milk, and grains like wheat, corn and oats; also, meat and eggs. Most of these foods really come to us out of the ground. Meat and eggs are made from the grain, grass, and other vegetables that the cattle and hens eat.

FAT-MAKING FOODS.

We need cushions and wrappings of fat, here and there in our bodies, to keep us

ho live
ey have
eep lick
em.
eek out
id go in

ith our
r when
l get a

esh; so
ese are
d oats;
foods
Meat
ss, and
ns eat.

of fat,
eep us

warm and make us comfortable. So we must have certain kinds of food that will make fat.

Esquimaux catching walrus.

There are right places and wrong places for fat, as well as for other things in this world. When alcohol puts fat into the muscles, that is fat badly made, and in the wrong place.

The good fat made for the parts of the

body which need it, comes from fat-making foods.

In cold weather, we need more fatty food than we do in summer, just as in cold countries people need such food all the time.

The Esquimaux, who live in the lands of snow and ice, catch a great many walrus and seal, and eat a great deal of fat meat. You would not be well unless you ate some fat or butter or oil.

WHAT WILL MAKE FAT?

Sugar will make fat, and so will starch, cream, rice, butter, and fat meat. As milk will make muscle and fat and bones, it is the best kind of food. Here again, it is the earth that sends us our food. Fat meat comes from animals well fed on grain and grass; sugar, from sugar-cane, maple-trees, or beets; oil, from olive-trees; butter, from cream; and starch, from potatoes, and from corn, rice, and other grains.

Green apples and other unripe fruits are not yet ready to be eaten. The starch which we take for food has to be changed into sugar, before it can mix with the blood and help feed the body. As the sun ripens fruit, it changes its starch to sugar. You can tell this by the difference in the taste of ripe and unripe apples.

CANDY.

Most children like candy so well, that they are in danger of eating more sugar than is good for them. You would starve if fed only on sugar.

W. would not need to be quite so much afraid of a little candy if it were not for the poison with which it is often colored.

Even wh. is called pure, white candy is sometimes not really such. There is a simple way by which you can find this out for yourselves.

If you put a spoonful of sugar into a tumbler of water, it will all dissolve and disappear. Put a piece of white candy into a tumbler of water; and, if it is made of pure sugar only, it will dissolve and disappear.

If it is not, you will find at the bottom of the tumbler some white earth. This is not

good food for anybody. Candy-makers often
put it into candy in place of sugar, because
it is cheaper than sugar.

REVIEW QUESTIONS.

1. Why do we need food?
2. How do people get water to drink?
3. Why is it not safe to drink water that has been standing in lead pipes?
4. Why is the water of a well that is near a drain or a stable, not fit to drink?
5. What food do the bones need?
6. How do we get lime for our bones?
7. What is said about salt?
8. What food do the muscles need?
9. Name some flesh-making foods.
10. Why do we need fat in our bodies?
11. What is said of the fat made by alcohol?
12. What kinds of food will make good fat?
13. What do the Esquimaux eat?
14. How does the sun change unripe fruits?
15. Why is colored candy often poisonous?
16. What is sometimes put into white candy? Why?
17. How could you show this?

CHAPTER XII.

HOW FOOD BECOMES PART OF THE BODY.

ERE, at last, is the bill of fare for our dinner:

Roast beef,	Bread,	Peaches,
Potatoes,	Butter,	Bananas,
Tomatoes,	Salt,	Oranges,
Squash,	Water,	Grapes.

What must be done first, with the different kinds of food that are to make up this dinner?

The meat, vegetables, and bread must be cooked. Cooking prepares them to be easily worked upon by the mouth and stomach. If they were not cooked, this work would be very hard. Instead of going on quietly and without letting us know any thing about it, there would be pains and aches in the over-worked stomach.

The fruit is not cooked by a fire; but we might almost say the sun had cooked it, for the sun has ripened and sweetened it.

When you are older, some of you may have charge of the cooking in your homes. You must then remember that food well cooked is worth twice as much as food poorly cooked.

"A good cook has more to do with the health of the family, than a good doctor."

THE SALIVA.

Next to the cooking comes the eating.

As soon as we begin to chew our food, a juice in the mouth, called saliva (sa lī′vä), moistens and mixes with it.

Saliva has the wonderful power of turning starch into sugar; and the starch in our food needs to be turned into sugar, before it can be taken into the blood.

You can prove for yourselves that saliva can turn starch into sugar. Chew slowly a piece of dry cracker. The cracker is made mostly of starch, because wheat is full of starch. At first, the cracker is dry and

; but we
ed it, for
).

you may
r homes.
ood well
)d poorly

with the
tor."

ing.
r food, a
(sa lī' vá),

of turn-
i in our
)efore it

t saliva
lowly a
is made
full of
ry and

tasteless. Soon, however, you find it tastes
sweet; the saliva is changing the starch into
sugar.

All your food should be eaten slowly and
chewed well, so that the saliva may be able
to mix with it. Otherwise, the starch may
not be changed; and if one part of your body
neglects its work, another part will have
more than its share to do. That is hardly
fair.

If you swallow your food in a hurry and
do not let the saliva do its work, the stomach
will have extra work. But it will find it
hard to do more than its own part, and,
perhaps, will complain.

It can not speak in words; but will
by aching, and that is almost as plain as
words.

SWALLOWING.

Next to the chewing, comes the swallowing. Is there any thing wonderful about
that?

We have two passages leading down our
throats. One is to the lungs, for breath-

ing; the other, to the stomach, for swallowing.

Do you wonder why the food does not sometimes go down the wrong way?

The windpipe leading to the lungs is in front of the other tube. It has at its top a little trap-door. This opens when we breathe and shuts when we swallow, so that the food slips over it safely into the passage behind, which leads to the stomach.

If you try to speak while you have food in your mouth, this little door has to open, and some bit of food may slip in. The windpipe will not pass it to the lungs, but tries to force it back. Then we say the food chokes us. If the windpipe can not succeed in forcing back the food, the person will die.

HOW THE FOOD IS CARRIED THROUGH THE BODY.

But we will suppose that the food of our dinner has gone safely down into the stomach. There the stomach works it over, and mixes in gastric juice, until it is all a gray fluid,

Now it is ready to go into the intestines,—
a long, coiled tube which leads out of the
stomach,—from which the prepared food is
taken into the blood.

The blood carries it to the heart. The
heart pumps it out with the blood into the
lungs, and then all through the body, to
make bone, and muscle, and skin, and hair,
and eyes, and brain.

Besides feeding all these parts, this dinner
can help to mend any parts that may be
broken.

Suppose a boy should break one of the
bones of his arm, how could it be mended?

If you should bind together the two parts
of a broken stick and leave them a while, do
you think they would grow together?

No, indeed!

But the doctor could carefully bind together the ends of the broken bone in the
boy's arm and leave it for a while, and the
blood would bring it bone food every day,
until it had grown together again.

So a dinner can both make and mend the
different parts of the body.

REVIEW QUESTIONS.

1. What shall we have for dinner?
2. What is the first thing to do to our food?
3. Why do we cook meat and vegetables?
4. Why do not ripe fruits need cooking?
5. What is said about a good cook?
6. What is the first thing to do after taking the food into your mouth?
7. Why must you chew it?
8. What does the saliva do to the food?
9. How can you prove that saliva turns starch into sugar?
10. What happens if the food is not chewed and mixed with the saliva?
11. What comes next to the chewing?
12. What is there wonderful about swallowing?
13. What must you be careful about, when you are swallowing?
14. What happens to the food after it is swallowed?
15. How is it changed in the stomach?
16. What carries the food to every part of the body?
17. How can food mend a bone?

CHAPTER XIII.

STRENGTH.

ERE are the names of some of the different kinds of food. If you write them on the blackboard or on your slates, it will help you to remember them.

Water. Salt. Lime.

Meat,
Milk,
Eggs,
Wheat,
Corn,
Oats, } for muscles.

Sugar,
Starch,
Fat,
Cream,
Oil, } for fat and heat.

Perhaps some of you noticed that we had no wine, beer, nor any drink that had alcohol in it, on our bill of fare for dinner. We had no cigars, either, to be smoked after dinner. If these are good things, we ought to have had them. Why did we leave them out?

We should eat in order to grow strong and keep strong.

STRENGTH OF BODY.

If you wanted to measure your strength, one way of doing so would be to fasten a heavy weight to one end of a rope and pass the rope over a pulley. Then you might take hold at the other end of the rope and pull as hard and steadily as you could, marking the place to which you raised the weight. By trying this once a week, or once a month, you could tell by the marks, whether you were gaining strength.

But how can we gain strength?

We must exercise in the open air, and take pure air into our lungs to help purify our blood, and plenty of exercise to make our muscles grow.

We must eat good and simple food, that the blood may have supplies to take to every part of the body.

ALCOHOL AND STRENGTH.

People used to think that alcohol made them strong.

Can alcohol make good muscles, or bone, or nerve, or brain?

You have already answered "No!" to each of these questions.

If it can not make muscles, nor bone, nor nerve, nor brain, it can not give you any strength.

BEER.

Some people may tell you that drinking beer will make you strong.

The grain from which the beer is made, would have given you strength. If you should measure your strength before and after drinking beer, you would find that you had not gained any. Most of the food part of the grain has been turned into alcohol.

CIDER.

The juice of crushed apples, you know, is called cider. As soon as the cider begins to turn sour, or "hard," as people say, alcohol begins to form in it.

Pure water is good, and apples are good. But the apple-juice begins to be a poison as soon as there is the least drop of alcohol in it. In cider-making, the alcohol forms in the juice, you know, in a few hours after it is pressed out of the apples.

None of the drinks in which there is alcohol, can give you real strength.

Then why do people think they can?

Because alcohol puts the nerves to sleep, they can not, truly, tell the brain how hard the work is, or how heavy the weight to be lifted.

The alcohol has in this way cheated men into thinking they can do more than they really can. This false feeling of strength lasts only a little while. When it has passed, men feel weaker than before.

A story which shows that alcohol does not give strength, was told me by the captain of a ship, who sailed to China and other distant places.

Many years ago, when people thought a little alcohol was good, it was the custom to carry in every ship, a great deal of rum. This liquor is distilled from molasses and contains about one half alcohol. This rum was given to the sailors every day to drink; and, if there was a great storm, and they had very

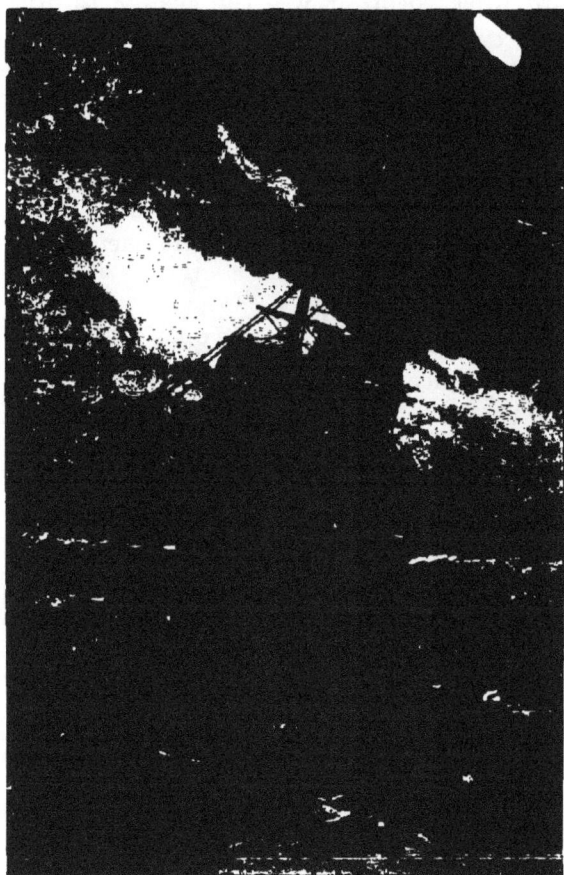

hard work to do, it was the custom to give them twice as much rum as usual.

The captain watched his men and saw that they were really made no stronger by drinking the rum; but that, after a little while, they felt weaker. So he determined to go to sea with no rum in his ship. Once out on the ocean, of course the men could not get any.

At first, they did not like it; but the captain was very careful to have their food good and plent... ; and, when a storm came, and they we... ...t and cold and tired, he gave them ho... ...offee to drink. By the time they had crossed the ocean, the men said: "The captain is right. We have worked better, and we feel stronger, for going without the rum."

STRENGTH OF MIND.

We have been talking about the strength of muscles; but the very best kind of strength we have is brain strength, or strength of mind.

Alcohol makes the head ache and deadens

om to give

1 and saw
tronger by
;er a little
ermined to
hip. Once
men could

,; but the
their food
orm came,
tired, he
y the time
men said :
orked bet-
g without

e stfength
f strength
rength of

.d deadens

the nerves, so that they can not carry their
messages correctly. Then the brain can not
think well. Alcohol does not strengthen the
mind.

Some people have little or no m   y, and
no houses or lands; but every person ought
to own a body and a mind that can work for
him, and make him useful and happy.

Suppose you have a strong, healthy body,
hands that are well-trained to work, and a
clear, thinking brain to be master of the whole.
Would you be willing to change places with a
man whose body and mind had been poisoned
by alcohol, tobacco, and opium, even though
he lived in a palace, and had a million of
dollars ?

If you want a mind that can study, understand, and think well, do not let alcohol and
tobacco have a chance to reach it.

REVIEW QUESTIONS.

1. What things were left out of our bill of fare?
2. How could you measure your strength?
3. How can you gain strength?
4. Why does drinking beer not make you strong?

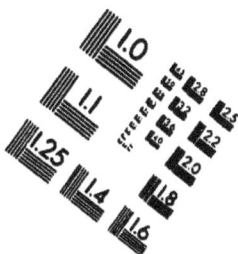

IMAGE EVALUATION
TEST TARGET (MT-3)

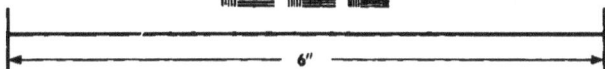

6"

Photographic
Sciences
Corporation

23 WEST MAIN STREET
WEBSTER, N.Y. 14580
(716) 872-4503

5. Show why drinking wine or any other alcoholic drink will not
   make you strong.
6. Why do people imagine that they feel strong after taking these
   drinks?
7. Tell the story which shows that alcohol does not help sailors
   do their work.
8. What is the best kind of strength to have?
9. How does alcohol affect the strength of the mind?

CHAPTER XIV.

THE HEART.

THE heart is in the chest, the upper part of the strong box which the ribs, spine, shoulder-blades, and collar-bones make for each of us.

It is made of very thick, strong muscles, as you can see by looking at a beef's heart, which is much like a man's, but larger.

HOW THE HEART WORKS.

Probably some of you have seen a fire-engine throwing a stream of water through a hose upon a burning building.

As the engine forces the water through the hose, so the heart, by the working of its strong muscles, pumps the blood through tubes, shaped like hose, which lead by thousands of little branches all through the body. These tubes are called arteries (är′tẽr iz).

Those tubes which bring the blood back again to the heart, are called veins (vänz). You can see some of the smaller veins in your wrist.

If you press your finger upon an artery in your wrist, you can feel the steady beating of the pulse. This tells just how fast the heart is pumping and the blood flowing.

The doctor feels your pulse when you are sick, to find out whether the heart is working too fast, or too slowly, or just right.

Some way is needed to send the gray fluid that is made from the food we eat and drink, to every part of the body.

To send the food with the blood :   sure way of making it reach every part.

So, when the stomach has prepared the food, the blood takes it up and carries it to every part of the body. It then leaves with each part, just what it needs.

THE BLOOD AND THE BRAIN.

As the brain has so much work to attend to, it must have very pure, good blood sent to it, to keep it strong. Good blood is made

ood back

anz). You

in your

artery in

eating of

the heart

you are

· working

:ray fluid

eat and

sure

ared the

·ies it to

ves with

N.

·o attend

ood sent

is made

from good food. It can not be good if it has been poisoned with alcohol or tobacco.

We must also remember that the brain needs a great deal of blood. If we take alcohol into our blood, much of it goes to the brain. There it affects the nerves, and makes a man lose control over his actions.

EXERCISE.

When you run, you can feel your heart beating. It gets an instant of rest between the beats.

Good exercise in the fresh air makes the heart work well and warms the body better than a fire could do.

DOES ALCOHOL DO ANY HARM TO THE HEART?

Your heart is made of muscle. You know what harm alcohol does to the muscles.

Could a fatty heart work as well as a muscular heart? No more than a fatty arm could do the work of a muscular arm. Besides, alcohol makes the heart beat too fast, and so it gets too tired.

REVIEW QUESTIONS.

1. Where is the heart placed?
2. Of what is it made?
3. What work does it do?
4. What are arteries and veins?
5. What does the pulse tell us?
6. How does the food we eat reach all parts of the body?
7. How does alcohol in the blood affect the brain?
8. When does the heart rest?
9. How does exercise in the fresh air help the heart?
10. What harm does alcohol do to the heart?

CHAPTER XV.

THE LUNGS.

THE blood flows all through the body, carrying good food to every part. It also gathers up from every part the worn-out matter that can no longer be used. By the time it is ready to be sent back by the veins, the blood is no longer pure and red. It is dull and bluish in color, because it is full of impurities.

If you look at the veins in your wrist, you will see that they look blue.

If all this bad blood goes back to the heart, will the heart have to pump out bad blood next time? No, for the heart has neighbors very near at hand, ready to change the bad blood to pure, red blood again.

THE LUNGS.

These neighbors are the lungs. They are

in the chest on each side of the heart. When
you breathe, their little air-cells swell out, or
expand, to take in the air Then they con-

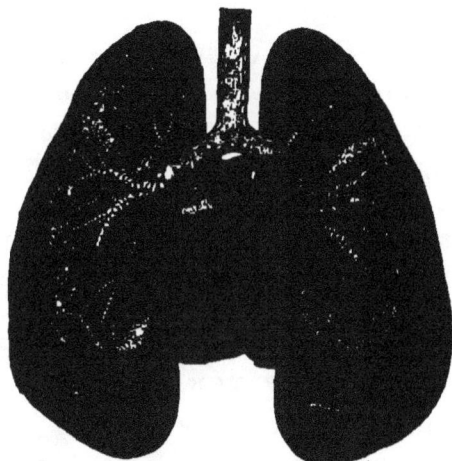

The lungs, heart, and air-passages.

tract again, and the air passes out through
your mouth or nose. The lungs must have
plenty of fresh air, and plenty of room to
work in.

If your clothes are too tight and the lungs
do not have room to expand, they can not
take in so much air as they should. Then

the blood can not be made pure, and the whole body will suffer.

For every good breath of fresh air, the lungs take in, they send out one of impure air.

In this way, by taking out what is bad, they prepare the blood to go back to the heart pure and red, and to be pumped out through the body again.

How the lungs can use the fresh air for doing this good work, you can not yet understand. By and by, when you are older, you will learn more about it.

CARE OF THE LUNGS.

Do the lungs ever rest?

You never stop breathing, not even in the night. But if you watch your own breathing you will notice a little pause between the breaths. Each pause is a rest. But the lungs are very steady workers, both by night and by day. The least we can do for them, is to give them fresh air and plenty of room to work in.

You may say: "We can't give them more

room than they have. They are shut up in
our chests."

I have seen people who wore such tight
clothes that their lungs did not have room
to take a full breath. If any part of the
lungs can not expand, it will become useless.
If your lungs can not take in air enough to
purify the blood, you can not be so well
and strong as God intended, and your life
will be shortened.

If some one was sewing for you, you would
not think of shutting her up in a little place
where she could not move her hands freely.
The lungs are breathing for you, and need
room enough to do their work.

THE AIR.

The lungs breathe out the waste matter
that they have taken from the blood. This
waste matter poisons the air. If we should
close all the doors and windows, and the fire-
place or opening into the chimney, and leave
not even a crack by which the fresh air could
come in, we would die simply from staying in
such a room. The lungs could not do their

up in

. tight
ɔ room
of the
ɪseless.
ɪgh to
o well
ur life

would
ɔ place
freely.
l need

matter
This
should
ɪe fire-
l leave
· could
ʳing in
· their

work for the blood, and the blood could not do its work for the body.

Impure air will poison you. You should not breathe it. If your head aches, and you feel dull and sleepy from being in a close room, a run in the fresh air will make you feel better.

The good, pure air makes your blood pure; and the blood then flows quickly through your whole body and refreshes every part.

We must be careful not to stay in close rooms in the day-time, nor sleep in close rooms at night. We must not keep out the fresh air that our bodies so much need.

It is better to breathe through the nose than through the mouth. You can soon learn to do so, if you try to keep your mouth shut when walking or running.

If you keep the mouth shut and breathe through the nose, the little hairs on the inside of the nose will catch the dust or other impurities that are floating in the air, and so save their going to the lungs. You will get out of breath less quickly when running if you keep your mouth shut.

DOES ALCOHOL DO ANY HARM TO THE LUNGS?

The little air-cells of the lungs have very delicate muscular (mŭs´ku lar) walls. Every time we breathe, these walls have to move. The muscles of the chest must also move, as you can all notice in yourselves, as you breathe.

All this muscular work, as well as that of the stomach and heart, is directed by the nerves.

You have learned already what alcohol will do to muscles and nerves, so you are ready to answer for stomach, for heart, and for lungs. Is alcohol a help to them?

REVIEW QUESTIONS.

1. Besides carrying food all over the body, what other work does the blood do?
2. Why does the blood in the veins look blue?
3. Where is the blood made pure and red again?
4. Where is it sent, from the lungs?
5. What must the lungs have in order to do this work?
6. When do the lungs rest?
7. Why should we not wear tight clothes?
8. How does the air in a room become spoiled?
9. How can we keep it fresh and pure?
10. How should we breathe?
11. Why is it better to breathe through the nose than through the mouth?
12. Why is alcohol not good for the lungs?

CHAPTER XVI.

THE SKIN.

THERE is another part of your body carrying away waste matter all the time—it is the skin.

The body is covered with skin. It is also lined with a more delicate kind of skin. You can see where the outside skin and the lining skin meet at your lips.

There is a thin outside layer of skin which we can pull off without hurting ourselves; but I advise you not to do so. Because under the outside skin is the true skin, which is so full of little nerves that it will feel the least touch as pain. When the outer skin, which protects it, is torn away, we must cover the true skin to keep it from harm.

In hot weather, or when any one has been **working** or playing hard, the face, and sometimes the whole body, is covered with little drops of water. We call these drops perspiration (pĕr spĭ ra'shŭn).

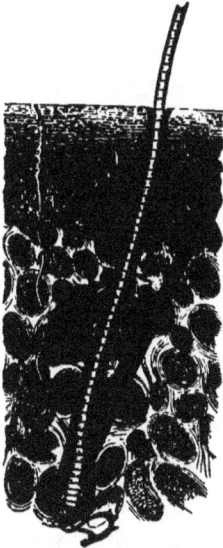

Where does it come from? It comes through many tiny holes in the skin, called pores (pōrz). Every pore is the mouth of a tiny tube which is carrying off waste matter and water from your body. If you could piece together all these little perspiration tubes that are in the skin of one person, they would make a line more than three miles long.

Perspiratory tube.

Sometimes, you can not see the perspiration, because there is not enough of it to form drops. But it is always coming out through your skin, both in winter and summer. Your body is kept healthy by having its worn-out matter carried off in this way, as well as in other ways.

THE NAILS.

The nails grow from the skin.

The finger nails are little shields to pro-
tect the ends of your fingers from getting
hurt. These finger ends are full of tiny
nerves, and would be badly off without such
shields. No one likes to see nails that have
been bitten.

CARE OF THE SKIN.

Waste matter is all the time passing out
through the perspiration tubes in the skin.
This waste matter must not be left to clog up
the little openings of the tubes. It should be
washed off with soap and water.

When children have been playing out-of-
doors, they often have very dirty hands and
faces. Any one can see, then, that they need
to be washed. But even if they had been in
the cleanest place all day and had not
touched any thing dirty, they would still
need the washing; for the waste matter that
comes from the inside of the body is just as
hurtful as the mud or dust of the street. You

do not see it so plainly, because it comes out very little at a time. Wash it off well, and your skin will be fresh and healthy, and able to do its work. If the skin could not do its work, you would die.

Do not keep on your rubber boots or shoes all through school-time. Rubber will not let the perspiration pass off, so the little pores get clogged and your feet begin to feel uncomfortable, or your head may ache. No part can fail to do its work without causing trouble to the rest of the body. But you should always wear rubbers out-of-doors when the ground is wet. Certainly, they are very useful then.

When you are out in the fresh air, you are giving the other parts of your body such a good chance to perspire, that your feet can bear a little shutting up. But as soon as you come into the house, take the rubbers off.

Now that you know what the skin is doing all the time, you will understand that the clothes worn next to your skin are full of little worn-out particles, brought out by the perspiration. When these clothes are

taken off at night, they should be so spread
out, that they will air well before morning. Never wear any of the clothes through
the night, that you have worn during the
day.

Do not roll up your night-dress in the
morning and put it under your pillow. Give
it first a good airing at the window and then
hang it where the air can reach it all day.
By so doing, you will have sweeter sleep at
night.

You are old enough to throw the bed-
clothes off from the bed, before leaving your
rooms in the morning. In this way, the bed
and bed-clothes may have a good airing. Be
sure to give them time enough for this.

WORK OF THE BODY.

You have now learned about four important kinds of work:—

1st. The stomach prepares the food for the
blood to take.

2d. The blood is pumped out of the heart
to carry food to every part of the body, and
to take away worn-out matter.

3d. The lungs use fresh air in making the dark, impure blood, bright and pure again.

4th. The skin carries away waste matter through the little perspiration tubes.

All this work goes on, day and night, without our needing to think about it at all; for messages are sent to the muscles by the nerves which keep them faithfully at work, whether we know it or not.

REVIEW QUESTIONS.

1. What covers the body?
2. What lines the body? .
3. Where are the nerves of the skin?
4. What is perspiration? What is the common name for it?
5. What are the pores of the skin?
6. How does the perspiration help to keep you well?
7. Of what use are the nails?
8. How should they be kept?
9. What care should be taken of the skin?
10. Why should you not wear rubber boots or overshoes in the house?
11. Why should you change under-clothing night and morning?
12. Where should the night-dress be placed in the morning?
13. What should be done with the bed-clothes? Why?
14. Name the four kinds of work about which you have learned.
15. How are the organs of the body kept at work?

CHAPTER XVII.

THE SENSES.

WE have five ways of learning about all things around us. We can see them, touch them, taste them, smell them, or hear them. Sight, touch, taste, smell, and hearing, are called the five senses.

You already know something about them, for you are using them all the time.

In this lesson, you will learn a little more about seeing and hearing.

THE EYES.

In the middle of your eye is a round, black spot, called the pupil. This pupil is only a hole with a muscle around it. When you are in the light, the muscle draws up, and makes the pupil small, because you can get all the light you need through a small opening. When you are in the dark, the muscle stretches, and opens the pupil wide to let in more light.

The pupils of the cat's eyes are very large in the dark. They want all the light they can get, to see if there are any mice about.

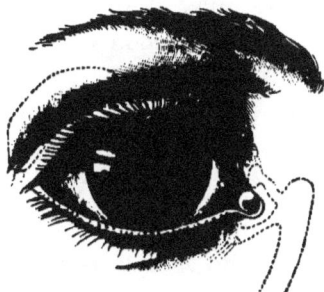

The pupil of the eye opens into a little, round room where the nerve of sight is. This is a

The eyelashes and the tear-glands.

safe place for this delicate nerve, which can not bear too much light. It carries to the brain an account of every thing we see.

We might say the eye is taking pictures for us all day long, and that the nerve of sight is describing these pictures to the brain.

CARE OF THE EYES.

The nerves of sight need great care, for they are very delicate.

Do not face a bright light when you are reading or studying. While writing, you should sit so that the light will come from the left side; then the shadow of your hand will not fall upon your work.

One or two true stories may help you to remember that you must take good care of your eyes.

The nerve of sight can not bear too bright a light. It asks to have the pupil made small, and even the eyelid curtains put down, when the light is too strong.

Once, there was a boy who said boastfully to his playmates: "Let us see which of us can look straight at the sun for the longest time."

Then they foolishly began to look at the sun. The delicate nerves of sight felt a sharp pain, and begged to have the pupils made as small as possible and the eyelid curtains put down.

But the foolish boys said "No." They were trying to see which would bear it the longest. Great harm was done to the brains as well as eyes of both these boys. The one who looked longest at the sun died in consequence of his foolish act.

The second story is about a little boy who tried to turn his eyes to imitate a schoolmate who was cross-eyed. He turned them; but he could not turn them back again. Although he is now a gentleman more than fifty years old and has had much painful work done upon his eyes, the doctors have never been able to set them quite right.

You see from the first story, that you must be careful not to give your eyes too much light. But you must also be sure to give them light enough.

When one tries to read in the twilight, the little nerve of sight says: "Give me more light; I am hurt, by trying to see in the dark."

If you should kill these delicate nerves, no others would ever grow in place of them, and you would never be able to see again.

THE EARS.

What you call your ears are only pieces of gristle, so curved as to catch the sounds and pass them along to the true ears. These are deeper in the head, where the nerve of hear-

oy who
school-
l them;
  again.
re than
painful
rs have
ght.
at you
yes too
sure to

vilight,
e more
in the

ves, no
m, and

eces of
ds and
se are
hear-

ing is waiting to send an account of each
sound to the brain.

CARE OF THE EARS.

The ear nerve is in less danger than that
of the eye. Careless children sometimes put
pins into their ears and so break the "drum."
That is a very bad thing to do. Use only a
soft towel in washing your ears. You should
never put any thing hard or sharp into them.

I must tell you a short ear story, about
my father, when he was a small boy.

One day, when playing on the floor, he
laid his ear to the crack of the door, to feel
the wind blow into it. He was so young that
he did not know it was wrong; but the next
day he had the earache severely. Although
he lived to be an old man, he often had
the earache. He thought it began from the
time when the wind blew into his ear from
under that door.

ALCOHOL AND THE SENSES.

All this fine work of touching, tasting, seeing, smelling, and hearing, is nerve work.

The man who is in the habit of using alcoholic drinks can not touch, taste, see, smell, or hear so well as he ought. His hands tremble, his speech is sometimes thick, and often he can not walk straight. Sometimes, he thinks he sees things when he does not, because his poor nerves are so confused by alcohol that they can not do their work.

Answer now for your taste, smell, and touch, and also for your sight and hearing; should their beautiful work be spoiled by alcohol?

REVIEW QUESTIONS.

1. Name the five senses.
2. What is the pupil of the eye?
3. How is it made larger or smaller?
4. Why does it change in size?
5. What can a cat's eyes do?
6. Where is the nerve of the eye?
7. What work does it do?
8. Why must one be careful of his eyes?
9. Where should the light be for reading or studying?
10. Tell the story of the boys who looked at the sun.
11. Tell the story of the boy who made himself cross-eyed.
12. Why should you not read in the twilight?
13. What would be the result, if you should kill the nerves of sight?
14. Where are the true ears?
15. How may the nerves of hearing be injured?
16. Tell the story of the boy who injured his ear.
17. How is the work of the senses affected by drinking liquor?

CHAPTER XVIII.

HEAT AND COLD.

WHAT MAKES US WARM?

"MY thick, warm clothes make me warm," says some child.

No! Your thick, warm clothes keep you warm. They do not make you warm.

Take a brisk run, and your blood will flow faster and you will be warm very quickly.

On a cold day, the teamster claps his hands and swings his arms to make his blood flow quickly and warm him.

Every child knows that he is warm inside; for if his fingers are cold, he puts them into his mouth to warm them.

If you should put a little thermometer into your mouth, or under your tongue, the mercury (mĕr′ku rў) would rise as high as it does out of doors on a hot, summer day.

This would be the same in summer or winter, in a warm country or a cold one, if you were well and the work of your body was going on steadily.

WHERE DOES THIS HEAT COME FROM?

Some of the work which is all the time going on inside your body, makes this heat.

The blood is thus warmed, and then it carries the heat to every part of the body. The faster the blood flows, the more heat it brings, and the warmer we feel.

In children, the heart pumps from eighty to ninety times a minute.

This is faster than it works in old people, and this is one reason why children are generally much warmer than old people.

But we are losing heat all the time.

You may breathe in cold air; but that which you breathe out is warm. A great deal of heat from your warm body is all the time passing off through your skin, into the cooler air about you. For this reason, a room full of people is much warmer than the same room when empty.

CLOTHING.

We put on clothes to keep in the heat which we already have, and to prevent the cold air from reaching our skins and carrying off too much heat in that way.

Most of you children are too young to choose what clothes you will wear. Others decide for you. You know, however, that woolen under-garments keep you warm in winter, and that thick boots and stockings should be worn in cold weather. Thin dresses or boots may look pretty; but they are not safe for winter wear, even at a party.

A healthy, happy child, dressed in clothes which are suitable for the season, is pleasanter to look at than one whose dress, though rich and handsome, is not warm enough for health or comfort.

When you feel cold, take exercise, if possible. This will make the hot blood flow all through your body and warm it. If you can not, you should put on more clothes, go to a warm room, in some way get warm and keep warm, or the cold will make you sick.

TAKING COLD.

If your skin is chilled, the tiny mouths of the perspiration tubes are sometimes closed and can not throw out the waste matter. Then, if one part fails to do its work, other parts must suffer. Perhaps the inside skin becomes inflamed, or the throat and lungs, and you have a cold, or a cough.

ALCOHOL AND COLD.

People used to think that nothing would warm one so well on a cold day, as a glass of whiskey, or other alcoholic drink.

It is true that, if a person drinks a little alcohol, he will feel a burning in the throat, and presently a glowing heat on the skin.

The alcohol has made the hot blood rush into the tiny tubes near the skin, and he thinks it has warmed him.

But if all this heat comes to the skin, the cold air has a chance to carry away more than usual. In a very little time, the drinker will be colder than before. Perhaps he will not know it; for the cheating alcohol

mouths
s closed
matter.
t, other
le skin
. lungs,

would
a glass

a little
throat,
skin.
od rush
and he

in, the
y more
e, the
Perhaps
alcohol

will have deadened his nerves so that they send no message to the brain. Then he may not have sense enough to put on more clothing and may freeze. He may even, if it is very cold, freeze to death.

People, who have not been drinking alcohol are sometimes frozen; but they would have frozen much quicker if they had drunk it.

Horse-car drivers and omnibus drivers have a hard time on a cold winter day. They are often cheated into thinking that alcohol will keep them warm; but doctors have learned that it is the water-drinkers who hold out best against the cold. Alcohol can not really keep a person warm.

All children are interested in stories about Arctic explorers, whose ships get frozen into great ice-fields, who travel on sledges drawn by dogs, and sometimes live in Esquimau huts, and drink oil, and eat walrus meat.

These men tell us that alcohol will not keep them warm, and you know why.

The hunters and trappers in the snowy regions of the Rocky Mountains say the same thing. Alcohol not only can not keep them warm; but it lessens their power to resist cold.

Scene in the Arctic regions.

Many of you have heard about the Greely party who were brought home from the Arctic seas, after they had been starving and freezing for many months.

There were twenty-six men in all. Of these, nineteen died. Seven were found alive by their rescuers; one of these died soon afterward. The first man who died, was the only one of the party who had ever been a drunkard.

Of the nineteen who died, all but one used tobacco. Of the six now living,—four never used tobacco at all; and the other two, very seldom.

The tobacco was no real help to them in time of trouble. It had probably weakened their stomachs, so that they could not make the best use of such poor food as they had.

REVIEW QUESTIONS.

1. Why do you wear thick clothes in cold weather?
2. How can you prove that you are warm inside?
3. What makes this heat?
4. What carries this heat through your body?
5. How rapidly does your heart beat?
6. How are you losing heat all the time?
7. How can you warm yourself without going to the fire?
8. Will alcohol make you warmer, or colder?
9. How does it cheat you into thinking that you will be warmer for drinking it?
10. What do the people who travel in very cold countries, tell us about the use of alcohol?
11. How did tobacco affect the men who went to the Arctic seas with Lieutenant Greely?

CHAPTER XIX.

WASTED MONEY.

COST OF ALCOHOL.

NOW that you have learned about your bodies, and what alcohol will do to them, you ought also to know that alcohol costs a great deal of money. Money spent for that which will do no good, but only harm, is certainly wasted, and worse than wasted.

If a boy or a girl save ten cents a week, it will take ten weeks to save a dollar.

You can all think of many good and pleasant ways to spend a dollar. What would the beer-drinker do with it? If he takes two mugs of beer a day, the dollar will be used up in ten days. But we ought not to say used, because that word will make us think it was spent usefully. We will say, instead, the dollar will be wasted, in ten days.

If he spends it for wine or whiskey, it will go sooner, as these cost more. If no money was spent for liquor in this country, people would not so often be sick, or poor, or bad, or wretched. We should not need so many policemen, and jails, and prisons, as we have now. If no liquor was drunk, men, women, and children would be better and happier.

COST OF TOBACCO.

Most of you have a little money of your own. Perhaps you earned a part, or the whole of it, yourselves. You are planning what to do with it, and that is a very pleasant kind of planning.

Do you think it would be wise to make a dollar bill into a tight little roll, light one end of it with a match, and then let it slowly burn up? That would be wasting it, you say!

Yes! it would be wasted, if thus burned. It would be worse than wasted, if, while burning, it should also hurt the person who held it. If you should buy cigars or tobacco with your dollar, and smoke them, you could soon burn up the dollar and hurt yourselves besides; yet in the Dominion of Canada alone, there are about fifteen million pounds of tobacco used every year. This is more than twenty tons for every day in the year. It would occupy nearly four years of school time to count the number of pounds of tobacco used in Canada in a single year, if one was counted every second of the time. If each pound of tobacco was valued at only fifty cents the cost of the tobacco used in Canada every year would be $7,500,000, or more than $20,000 each day.

REVIEW QUESTIONS.

1. How may one waste money?
2. Name some good ways for spending money.
3. How does the liquor-drinker spend his money?
4. What could we do, if no money was spent for liquor?
5. Tell two ways in which you could burn up a dollar bill.
6. Which would be the safer way?
7. How much money is spent for tobacco, yearly, in this country?

INDEX.